LIFTOFF! SPACE EXPLORATION

DESTINATION MARS

BY NATHAN MILOSZEWSKI

Gareth Stevens PUBLISHING

Please visit our website, www.garethstevens.com. For a free color catalog of all our high-quality books, call toll free 1-800-542-2595 or fax 1-877-542-2596.

Library of Congress Cataloging-in-Publication Data

Names: Miloszewski, Nathan, author.
Title: Destination Mars / Nathan Miloszewski.
Description: New York : Gareth Stevens Publishing, [2021] | Series: Liftoff! space exploration | Includes index. | Contents: Discover the red planet -- Early observations -- Paving the way for the future -- The first close-up look at Mars -- Volcanoes, canyons, moons, and more -- 20 seconds on Mars -- Life on Mars? -- Mars used to be warm and wet -- The most powerful camera -- Next generation rovers -- Sending humans to Mars -- Why Mars? -- The future of Mars exploration.
Identifiers: LCCN 2019060251 | ISBN 9781538258729 (library binding) | ISBN 9781538258705 (paperback) | ISBN 9781538258712 (6 Pack) | ISBN 9781538258736 (ebook)
Subjects: LCSH: Mars (Planet)--Juvenile literature. | Mars (Planet)--Exploration--Juvenile literature.
Classification: LCC QB641 .M543 2021 | DDC 559.9/23--dc23
LC record available at https://lccn.loc.gov/2019060251

First Edition

Published in 2021 by
Gareth Stevens Publishing
111 East 14th Street, Suite 349
New York, NY 10003

Designer: Katelyn E. Reynolds
Editor: Abby Badach Doyle

Photo credits: Cover, p. 1 Vadim Sadovski/Shutterstock.com; cover, pp. 1-32 (series background) David M. Schrader/Shutterstock.com; p. 5 NASA's Goddard Space Flight Center; p. 7 Christos Georghiou/Shutterstock.com; p. 9 Universal History Archive/Getty Images; p. 10 NASA/JPL-Caltech/Dan Goods; p. 11 SSPL/Getty Images; pp. 13 (inset), 17 (inset) NASA / JPL; p. 13 (main) NASA/JPL-Caltech/GSFC/Univ. of Arizona; p. 15 (top and middle) Sovfoto/Universal Images Group via Getty Images; p. 15 (bottom) NASA/Wingman4l7/Wikipedia.org; p. 17 (main) The LIFE Picture Collection via Getty Images; pp. 19 (main), 23 (both), 25 NASA/JPL-Caltech; p. 19 (inset) NASA/SCIENCE PHOTO LIBRARY/Getty Images Plus; p. 21 (main) NASA; p. 21 (inset) NASA/JPL-Caltech/Univ. of Arizona; p. 27 © iStockphoto.com/e71lena; p. 29 (main) Dima Zel/Shutterstock.com; p. 29 (inset) Kevork Djansezian/Getty Images.

Printed in the United States of America

CONTENTS

WORDS IN THE GLOSSARY APPEAR IN **BOLD** TYPE THE FIRST TIME THEY ARE USED IN THE TEXT.

DISCOVER THE RED PLANET

We don't know who first discovered Mars because it's one of five planets that can be seen from Earth without a telescope. Humans have tracked it ever since we started looking at the night sky. Someone first saw this reddish "star" and gave it the nickname the Red Planet.

In 1609, Italian inventor Galileo Galilei built a telescope. He became the first person to use a telescope for **astronomy**. It's believed he was the first person to study Mars through a magnified lens. In the 1700s, early observations of the planet's surface uncovered dark spots, white spots, and straight lines. Some people thought the lines were canals dug by other living beings. The idea of life on Mars was born long ago. It still inspires our ongoing exploration.

INTERSTELLAR INFORMATION

ANCIENT PEOPLE THOUGHT MARS'S COLOR CAME FROM THE BLOOD OF SOLDIERS IN BATTLE. THEY NAMED IT MARS, AFTER THE ROMAN GOD OF WAR. THE RED ACTUALLY COMES FROM RUST PARTICLES IN THE SOIL, CALLED IRON OXIDE.

ON THE LEFT, AN ARTIST SHOWS HOW MARS MAY HAVE LOOKED WHEN SCIENTISTS BELIEVE IT HAD LIQUID WATER. AT RIGHT IS MARS TODAY: RED, DRY, AND DUSTY.

EARTH'S COUSIN

Billions of years ago, Mars could have been considered Earth's cousin because of how much the two planets looked like each other. Research points to evidence of liquid water on the surface of Mars that could have been in the form of rivers, lakes, and maybe even an ocean. Methane gas in the atmosphere has also been detected. On Earth, methane comes from animals such as cows, so it's possible that Mars used to be able to support life.

EARLY OBSERVATIONS

The early Babylonians, Egyptians, Greeks, and Romans all recognized a similar pattern when they studied the skies. Around 400 BCE, ancient people started to notice that there were five things in the sky that were very bright and moved differently than the other stars. They tracked their movement to **synchronize** their calendars, mark religious events, and navigate over land and sea.

What they were seeing were the planets that you can see from Earth without a telescope. In order of distance from the sun, those planets are Mercury, Venus, Mars, Jupiter, and Saturn. While they may not have fully understood what they were witnessing, they did recognize that the stars looked like they were fixed in the sky when compared to how the sun moved.

INTERSTELLAR INFORMATION

THIS MEMORY DEVICE CAN HELP YOU REMEMBER THE ORDER OF THE PLANETS, FROM CLOSEST TO FURTHEST: "MY VERY EDUCATED MOTHER JUST SERVED US NACHOS," OR MERCURY, VENUS, EARTH, MARS, JUPITER, SATURN, URANUS, AND NEPTUNE.

MARS IS ABOUT 140 MILLION MILES (225 MILLION KM) FROM EARTH. THE ROBOTS SENT FROM EARTH IN MISSIONS TO MARS HAVE TAKEN ABOUT EIGHT MONTHS TO REACH ITS SURFACE.

THE WANDERERS

More than 2,000 years ago, the Greeks noticed "fixed stars" and "wandering stars." These "wanderers" were the five moving planets they could see, plus the sun and the moon. That's where we get the word planet – it comes from the ancient Greek word for wanderer. The Babylonians are credited with giving us the seven-day week. They measured time by the movement of these seven **celestial** bodies, with a weekday named after each one.

PAVING THE WAY FOR THE FUTURE

The 1600s were not kind to astronomy. Galileo was famously **interrogated** by the Catholic church. He was found guilty during the Inquisition, a period of several hundred years when religion was trying to fight **heresy**. For example, the church believed that the solar system revolved around Earth, while research showed that the solar system revolved around the sun. Galileo was put in prison for his nontraditional views, but a book he wrote survived. That book became the foundation of modern physics.

Scientists in the 1700s had more freedom. During this time, they made discoveries with the telescope that helped lead to future Mars research. Even with their limited technology, some of the findings of this time were confirmed hundreds of years later with modern methods.

INTERSTELLAR INFORMATION

EARTH'S ATMOSPHERE IS 78 PERCENT NITROGEN AND 21 PERCENT OXYGEN, WITH SMALL AMOUNTS OF OTHER GASES SUCH AS ARGON AND CARBON DIOXIDE. THE ATMOSPHERE ON MARS IS 95 PERCENT CARBON DIOXIDE, 3 PERCENT NITROGEN, 1.6 PERCENT ARGON, AND ONLY SMALL AMOUNTS OF OXYGEN.

USING A TELESCOPE, GALILEO GALILEI DISCOVERED THAT OUR SOLAR SYSTEM REVOLVES AROUND THE SUN. THIS IS CALLED THE HELIOCENTRIC THEORY.

IMPORTANT DISCOVERIES

Even without modern technology, early astronomers made some surprisingly **accurate** observations about Mars.

1719
Giancomo Miraldi correctly assumes that white spots on Mars's North and South Poles are ice caps.

1784
Sir William Herschel correctly reasons that Mars has a thin atmosphere when he sees faint stars pass close by and their brightness doesn't change.

1840
Wilhelm Beer and Johann von Maedler correctly determine the length of a Martian day within 0.1 second of today's accepted time.

THE FIRST CLOSE-UP LOOK AT MARS

It took until the 20th century for humans to get an up-close look at Mars. The National Aeronautics and Space Administration (NASA), an independent agency of the United States federal government, wasn't successful at first. Mariner 3 was launched in 1964, but the rocket failed and didn't make it to Mars. On November 28, 1964, NASA launched Mariner 4, an **identical** unmanned spacecraft.

About eight months later on July 14, 1965, it made history in its flyby of Mars, taking the first photographs of another planet from space. The pictures showed impact craters, similar to what was being photographed up close from the moon. Mariner 4 was expected to survive for less than a year. Instead, it lasted almost three years!

MARINER 4'S FIRST PHOTO OF MARS, 1965

INTERSTELLAR INFORMATION

MARINER 4 SENT 21 PICTURES BACK TO EARTH AT A RATE OF 33.3 BITS PER SECOND THROUGH ITS ANTENNA. NASA'S ROVERS THAT ARE CURRENTLY ON THE SURFACE OF MARS CAN SEND DATA MUCH QUICKER, AT 11,000 BITS PER SECOND.

MARINER 4 WAS CONSIDERED A GREAT SUCCESS. IT SENT VALUABLE INFORMATION BACK TO EARTH THAT WOULD HELP FUTURE MISSIONS, SUCH AS DATA ABOUT SOLAR WINDS.

WHAT DID MARINER 4 FIND?

Mariner 4 lasted a lot longer than it was expected to, which provided a lot more information to scientists about Mars. Mariner 4 was able to coordinate measurements with Mariner 5, another spacecraft on its way to explore Venus. Mariner 4 used solar panels to power its scientific equipment, which included cameras, probes, a telescope, and various detectors. The images and data sent back to Earth showed that there was no visible water or signs of life on Mars.

VOLCANOES, CANYONS, MOONS, AND MORE

Successful Mariner missions to Mars continued through the 1960s and 1970s with Mariner 6, 7, and 9. Scientists took more pictures and continued to study the atmosphere. Some discoveries had to wait because the entire planet was covered in a monthlong dust storm. Mariner 9 sat in orbit until it cleared.

Since the 1800s, scientists thought they saw canals on Mars's surface which could be a sign of intelligent life. Images from the Mariner explorations proved that to be untrue—they were simply streaks of dust. The Martian moons, Phobos and Deimos, were first discovered in 1877 by Asaph Hall. However, in 1971, Mariner 9 took the first detailed photos of them. Like Earth's moon, they have craters. Unlike Earth's moon, they are irregularly shaped. Scientists believe they may be captured asteroids.

INTERSTELLAR INFORMATION

STUDYING MARS'S MOONS MIGHT HELP SCIENTISTS UNDERSTAND HOW EARTH WAS FORMED AND ITS SOURCE OF WATER. MARS'S MOONS MAY HAVE COME FROM A NEARBY ASTEROID BELT, AND THEN BECAME CAPTURED BY THE GRAVITY ON MARS. EARTH'S MOON IS THOUGHT TO BE DEBRIS CREATED FROM A LARGE OBJECT COLLIDING WITH EARTH.

PHOBOS AND DEIMOS ARE AMONG THE SMALLEST MOONS IN THE SOLAR SYSTEM. PHOBOS MAKES THREE ORBITS IN ONE DAY. DEIMOS IS FURTHER AWAY AND TAKES 30 HOURS FOR ONE ORBIT.

A CLOSER LOOK AT THE LANDSCAPE

Mariner 9 greatly expanded what we know about Mars. It was able to map the entire surface with photos. This time, the images showed some surprises! They found giant volcanoes, a canyon 3,000 miles (4,828 km) long, evidence of dried riverbeds, and a crater 6 miles (9.7 km) wide on the moon Phobos. Since this mission, scientists have talked about the idea of using the Martian moons as a base for astronauts.

20 SECONDS ON MARS

In 1971, the Soviet Union became the first country to land a man-made object on Mars. The Soviets' Mars 2 and Mars 3 spacecraft each had two parts: a lander for exploring the surface and an orbiter to circle Mars from above. The Mars 2 orbiter was successful, but the lander crashed. Mars 3 used parachutes, **aerobraking**, and special rockets for a soft landing. On December 2, 1971, Mars 3 reached the surface and took a picture. It survived for 20 seconds before losing communication.

Both landers carried a tiny rover called the PROP-M to test the soil. It could sense obstacles and back up, but it could only travel about 50 feet (15 m). These were the first rovers to land on Mars, although they never got the chance to explore.

INTERSTELLAR INFORMATION

A SOFT LANDING IS EXACTLY WHAT IT SOUNDS LIKE. A SPACECRAFT USES BRAKING **TECHNIQUES** AND TECHNOLOGY TO MAKE A CONTROLLED DESCENT WITHOUT CAUSING DAMAGE. THE OPPOSITE IS A CRASH LANDING, OR HITTING THE GROUND TOO HARD.

MARS 2 AND MARS 3 HAD FOUR PETALS THAT COULD OPEN AFTER LANDING TO STABILIZE THE CRAFT. THEN, INSTRUMENTS COULD TAKE READINGS AND SEND DATA BACK TO EARTH.

MARS 3 ASSEMBLY (SOLAR PANELS EXTENDED AND PARACHUTE CONTAINER ON TOP)

MARS 3 ASSEMBLY (LANDING CAPSULE WITH PARACHUTE CONTAINER ON TOP)

PROP-M ROVER

A WIN FOR THE SOVIETS

While the Mars 2 and Mars 3 missions didn't go as planned, these spacecraft achieved a huge series of firsts for the Soviets during the time of the Space Race rivalry. In the 20 seconds that the Mars 3 lander survived, it was able to snap the first photo from Mars's surface and send it back. Although the PROP-M rover wasn't able to be used, it landed on Mars 26 years before NASA's Sojourner rover.

LIFE ON MARS?

On July 20, 1976, Viking 1 became the first U.S. mission to land on Mars and take clear images of the surface. The Viking 2 lander touched down weeks later to explore another part of the planet. In prior Mariner missions, there was some disappointment about what was found on Mars. At that time, the data suggested Mars was a dead planet with no signs of life. Now, with two Viking landers combing the surface, NASA could test the soil.

Scientists didn't find the answers people were looking for. Biology experiments found some chemical activity in the soil that they didn't expect. Still, it wasn't clear evidence of any living microorganisms. This didn't end the search for life. There may have been life on Mars in the distant past.

INTERSTELLAR INFORMATION

VIKING 1 LASTED THE LONGEST, FROM 1976 TO 1982. IT TOOK MORE THAN 1,400 IMAGES OF VOLCANOES, LAVA PLAINS, HUGE CANYONS, CRATERS, AND SOME EVIDENCE OF ANCIENT OR FROZEN SURFACE WATER.

MODEL OF THE VIKING 1 AND VIKING 2 LANDERS

THE PHOTO FROM MARS 3 WAS TOO BLURRY TO SEE MUCH. IN THIS IMAGE FROM VIKING, YOU CAN SEE ROCKS, WINDBLOWN SAND OR DUST, AND THE FOOTPAD OF THE SPACECRAFT (FAR RIGHT).

STERILE SOIL

Scientists believe that a dry environment and intense ultraviolet radiation from the atmosphere might make Martian soil **sterile**. These conditions could prevent anything from living, growing, or even forming on Mars. On the Viking mission, they found that the planet's surface is a type of iron-rich clay that releases oxygen when it gets wet. Water vapor was found near the north polar cap, which is water ice. Mars's southern ice cap is dry ice from carbon dioxide.

WHAT PATHFINDER FOUND

It took 20 years for NASA to launch another spacecraft to Mars. The Pathfinder lander reached Mars on July 4, 1997. It delivered the Sojourner rover, the first successful robotic rover to study and explore the surface of Mars.

The rover wasn't very big. It only weighed 23 pounds (10.4 kg). However, it carried many instruments to send back lots of scientific data. These instruments were able to study the Martian atmosphere, climate changes, and test what types of materials made up the rocks and soil on Mars. Scientists found evidence that Mars was probably once warmer and wetter than it is now. The landing site NASA chose was an ancient floodplain with rocks like pebbles and cobbles, or rocks on Earth that have been rounded by running water.

INTERSTELLAR INFORMATION

PATHFINDER HAD A VERY CREATIVE SOFT LANDING. MANY AIRBAGS SURROUNDED THE ENTIRE SPACECRAFT. THEN, IT TUMBLED LIKE A GIANT BALL! IT BOUNCED 15 TIMES AND AS HIGH AS 50 FEET (15 M) BEFORE COMING TO REST UNHARMED.

SOJOURNER WAS ONLY EXPECTED TO LAST A WEEK, BUT IT SPENT 83 DAYS TAKING PICTURES AND OTHER MEASUREMENTS AS IT ROLLED AROUND MARS.

SIX WHEELS ARE BETTER THAN FOUR!

Mars is rocky, so NASA needed to design a rover that could handle uneven terrain. Instead of using springs like you'd find in a regular car, they used rocker arms and joints. These could **conform** to whatever surface Sojourner traveled on. Instead of four wheels, Sojourner rolled around on six. With two extra wheels, Sojourner could climb obstacles that were three times larger than it could if it only had four wheels!

THE MOST POWERFUL CAMERA

In 2005, NASA sent an extremely powerful camera to orbit Mars. It's known as the Mars **Reconnaissance** (ruh-KAH-nuh-sihns) Orbiter, or MRO for short. It can capture better details than ever of the planet's surface, terrain, and geological formations. The MRO also has a sounder to search for water under the surface of Mars.

According to NASA, other cameras sent to Mars could identify an object about the size of a school bus. The MRO's camera can identify an object the size of a dinner table. Landing a robot on Mars is difficult and dangerous. Being able to see any obstacles helps to keep equipment safe and not waste a mission. With the ability to locate water, scientists can target better landing sites.

INTERSTELLAR INFORMATION

WHY DO SCIENTISTS CARE ABOUT FINDING WATER? STUDYING WATER HELPS US UNDERSTAND IF MARS EVER SUPPORTED LIFE, IF HUMANS COULD TRAVEL THERE, AND HOW CLIMATE CHANGE MIGHT HAVE AFFECTED THE PLANET'S WATER SUPPLY.

THE MRO'S PRIMARY MISSION WAS SUPPOSED TO LAST TWO YEARS. LIKE OTHER SPACECRAFT IT LASTED LONGER THAN EXPECTED. IT HAS ORBITED MARS 60,000 TIMES OVER 13 YEARS.

PICTURE FROM THE MARS RECONNAISSANCE ORBITER (MRO) ▶

THE SEARCH FOR WATER

Scientists use the Mars Reconnaissance Orbiter's camera to zoom in on any sources or historical sources of water. That could be water vapor, ice, or liquid water. Using radar, scientists can "see" about a third of a mile (0.5 km) below the surface. There, they can get information about all the different layers of rock, ice, and maybe even water. They also study the insides of the Martian ice caps. The sounder measures water vapor, dust, and temperature in the atmosphere.

NEXT GENERATION ROVERS

As scientists continue to learn about Mars and how to send spacecraft there, the technology has gotten better over time. In January 2004, two rovers named Spirit and Opportunity landed on Mars in areas that looked like they had water in the past. Their main goal was to test a variety of rocks and soils to get clues. Spirit lasted until 2011. Opportunity lasted almost 15 years, until 2019. It ended communication after a large dust storm.

On August 5, 2012, NASA's largest rover ever landed on Mars. Instead of using a system of airbags, the Curiosity rover landed on its wheels. Curiosity is about the size of a car, can climb over knee-high obstacles, and can drill into the soil. It even has a laser that can vaporize rocks and study them.

INTERSTELLAR INFORMATION

CURIOSITY'S LASER CAN ZAP TARGETS UP TO 23 FEET (7 M) AWAY. THE LASER CREATES A SPARK THAT CONTAINS DIFFERENT GASES. THEN, THE ROVER STUDIES THAT SPARK TO SEE WHAT CHEMICALS ARE PRESENT.

ROVERS HAVE GOTTEN BIGGER AND MORE POWERFUL OVER THE YEARS. THE SMALLEST ROVER SHOWN HERE IS SOJOURNER FROM 1997. SPIRIT AND OPPORTUNITY WERE MIDSIZED. CURIOSITY IS THE BIGGEST.

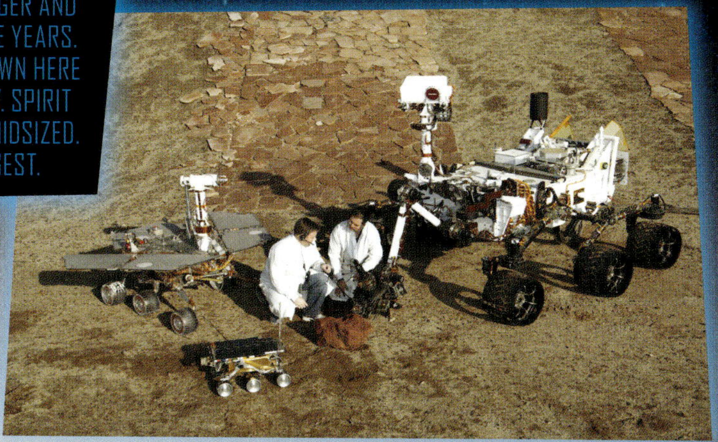

A MOBILE LABORATORY

The Curiosity rover is so big and so advanced that it's like a mobile laboratory. Curiosity is longer, heavier, and carries more equipment and cameras than the rovers before it. Like previous rovers, it still has six wheels and the same suspension system. It has a robotic arm 7 feet (2.1 m) long that helps it get close to rocks to study them. Curiosity was the first rover to get samples from drilled rocks on Mars.

SENDING HUMANS TO MARS

NASA's latest mission to Mars builds on the failures and successes of those before it. All of the past research, data, and discoveries will be used to find out if it's possible to send humans to Mars. In the Mars 2020 mission, NASA and the European Space Agency (ESA) are working together to send two rovers to Mars and test for ancient signs of life.

The Mars 2020 rover is expected to drill for core rock and soil samples that it will set aside for another mission to bring them back to Earth. If that succeeds, scientists will be able to study Mars from Earth, using equipment that is too large to send to Mars. This rover will also test a way to make oxygen from the Martian atmosphere. The Mars 2020 Rover mission is expected to last at least one Martian year. In Earth time, that's 687 days.

INTERSTELLAR INFORMATION

IN THE MARS 2020 MISSION, SCIENTISTS WILL ALSO CONTINUE TO STUDY WEATHER PATTERNS TO SEE IF THEY MAKE ACCURATE PREDICTIONS. THIS CAN HELP KEEP SPACECRAFT SAFE AND NOT GET STUCK IN STORMS OR EXTREME TEMPERATURES.

DIFFERENCES BETWEEN MARS 2020 AND CURIOSITY

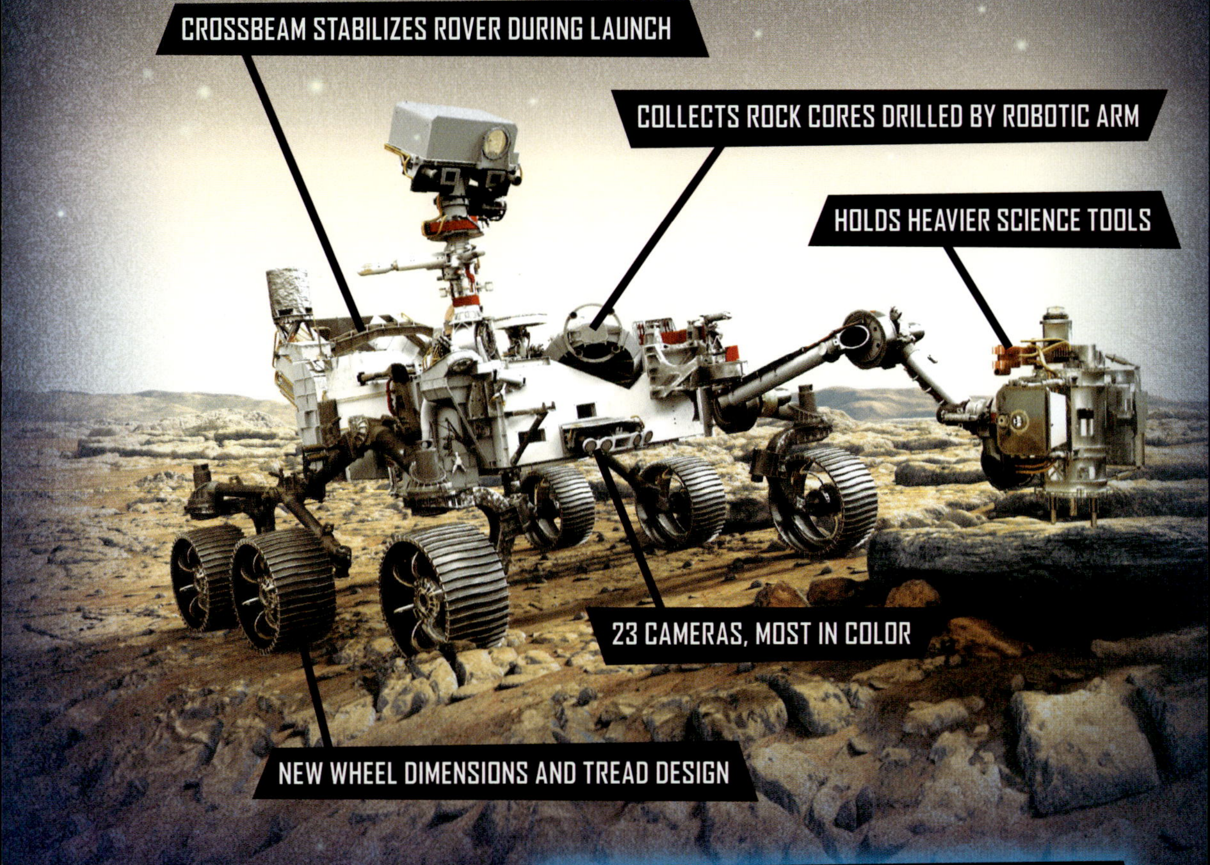

MARS 2020 LOOKS SIMILAR TO CURIOSITY, BUT IT HAS MORE ADVANCED EQUIPMENT. IT ALSO IS HEAVIER THAN CURIOSITY.

LIVING AND WORKING ON MARS

Launching heavy rockets into space is expensive and difficult. There's only so much equipment that can be sent to Mars at any one time. The current approach is to find ways to use the natural resources on Mars to make things humans need while they're there, especially water and oxygen. The atmosphere on Mars is 95 percent carbon dioxide. Finding a way to support basic life functions, such as having oxygen to breathe, would be a big step.

WHY MARS?

Why do we keep sending rovers and satellites to study this planet? Mars is the one planet in the solar system that is a bit like Earth and could possibly support human trips there. That's just not the case with other planets.

Scientists have found that Earth's moon and Mercury are both dry and have no air. Venus traps heat like a greenhouse. It has a lot of carbon dioxide that has made all the water escape. Plus, the temperatures on Venus are hotter than 900°F (500°C)! Mars has an atmosphere, ice on its polar caps, and may have water under the surface that could be mined. Spacecraft can also travel to Mars in a reasonable amount of time.

INTERSTELLAR INFORMATION

THE OLDEST ROCKS ON EARTH SHOW SIGNS OF LIFE FROM MORE THAN 4 BILLION YEARS AGO. SURFACES ON MARS THAT ARE ABOUT THAT AGE SHOW SIGNS OF WATER THAT USED TO BE THERE, SUCH AS ANCIENT LAKES.

IN 2016, NASA RECEIVED MORE THAN 18,000 APPLICATIONS TO ITS ASTRONAUT PROGRAM. IT WAS THEIR HIGHEST AMOUNT EVER! COULD ONE OF THEM BE THE FIRST HUMAN ON MARS?

ARE WE ALONE IN THE UNIVERSE?

There are things about Mars and Earth that are so similar that it makes scientists wonder if Mars had at one time been like Earth. If so, why did it change? For those reasons, Mars is the best planet to try to explore where life came from. It's also a living experiment. Could changes on Earth ever cause it to become more like Mars? If so, scientists want to find a way to fix that.

THE FUTURE OF MARS EXPLORATION

More countries and private companies around the world are trying to reach Mars now than at any other time in the history of space exploration. The best time to launch any spacecraft to Mars comes around every 26 months or so. This creates the shortest trip because it is when Earth and Mars are closest together. The most recent time for this launch window was the year 2020.

Groups from Europe, China, India, United Arab Emirates (UAE), NASA, and more are all competing for the same airspace now. More missions mean more opportunity for research. However, it might also create conflict or traffic jams. This will be a good test to see how countries will respect each other in space.

INTERSTELLAR INFORMATION

SCIENTISTS IN EUROPE AND THE UNITED STATES ARE WORKING ON A ROBOT SPACESHIP TO COLLECT SAMPLES OF MARTIAN ROCKS AND DELIVER THEM BACK TO EARTH. THIS MAY BE OUR BEST CHANCE YET TO FIND OUT IF LIFE EXISTED ON MARS AND WHETHER IT CAN SUPPORT HUMAN EXPLORATION.

ELON MUSK, THE FOUNDER AND CEO OF SPACEX, IS VOCAL ABOUT HIS COMPANY'S PLANS TO EXPLORE MARS. HE THINKS HUMANS COULD ESTABLISH A CITY THERE AS EARLY AS 2050.

WHEN WILL HUMANS GO TO MARS?

Getting humans on Mars feels closer to reality now, but there are still serious things to work out first. We still need to develop technology to get drinkable water from Mars, produce oxygen, and protect astronauts from high levels of radiation. Earth's atmosphere protects us from the harmful effects of the sun. On Mars, though, a person could develop cancer from radiation doses that are 700 times higher than they are on Earth.

GLOSSARY

accurate: free from mistakes

aerobraking: a type of braking a spacecraft uses to create drag against the atmosphere to slow down

astronomy: the scientific study of stars, planets, and other objects in outer space

celestial: of or relating to the sky

conform: to be similar to or the same as something

heresy: a belief or opinion that does not agree with the official belief or opinion of a particular religion

identical: being the same

interrogate: to ask someone questions in a thorough and often forceful way

reconnaissance: an activity in which something is sent out to find information

sterile: not able to produce crops or plants

synchronize: to cause things to agree in time

technique: a particular skill or ability to perform a job

FOR MORE INFORMATION

BOOKS

Aldrin, Buzz and Marianne Dyson. *Welcome to Mars: Making a Home on the Red Planet.* Washington, DC: National Geographic Children's Books, 2015.

Berger, Melvin. *Discovering Mars: The Amazing Story of the Red Planet.* New York, NY: Scholastic, 2015.

Motum, Markus. *Curiosity: The Story of a Mars Rover.* Somerville, MA: Candlewick, 2018.

WEBSITES

Distance to Mars
www.distancetomars.com/
Take a fun journey to see how long a trip to Mars would take if Earth were 100 pixels wide.

NASA: Mars Facts
mars.nasa.gov/all-about-mars/facts/
How long is a day on Mars? How long is a year? What is the average temperature on Mars? Find out answers to all these questions and more.

NASA Science Space Place
spaceplace.nasa.gov/menu/space/
Explore the rest of the galaxy with games and other activities on this interactive site.

Publisher's note to educators and parents: Our editors have carefully reviewed these websites to ensure that they are suitable for students. Many websites change frequently, however, and we cannot guarantee that a site's future contents will continue to meet our high standards of quality and educational value. Be advised that students should be closely supervised whenever they access the internet.

INDEX